A QUABBIN FARM ALBUM

a *Quabbin farm album*

Cathy Stanton

featuring photographs by **Oliver Scott Snure**

Copyright 2018 by Cathy Stanton.

Copyright 2018 by Oliver Scott Snure.

Also copyright 2018 by David Brothers, Anne Diemand Bucci, Jeannette Fellows, Cristina Garcia, Laura Moore, Julie Rawson, and Nina Wellen.

All rights reserved. With the exception of short excerpts in a review or critical article, no part of this book may be re-produced by any means, including information storage and retrieval or photocopying equipment, without written permission of the publisher, Haley's.

Haley's
488 South Main Street
Athol, MA 01331
haley.antique@verizon.net
800.215.8805

Cover designed by Cathy Stanton and Marcia Gagliardi.

Copy edited by Mary-Ann DeVita Palmieri.

Recipes compiled by Robin Shtulman.

Cataloguing in Publishing data:
Names: Stanton, Cathy, author.
Title: A Quabbin farm album / Cathy Stanton ; featuring photographs by Oliver Scott Snure.
Description: Athol, MA : Haley's, 2017.
Identifiers: LCCN 2017024017 | ISBN 9780996773041
Subjects: LCSH: Farms--Massachusetts. | Local foods--Massachusetts.
Classification: LCC S561.6.M4 S73 2017 | DDC 636/.0109744--dc23
LC record available at https://lccn.loc.gov/2017024017

*No matter how much one may love the world as a whole,
one can live fully in it only by
living responsibly in some small part of it.*
—Wendell Berry
The Art of the Commonplace: The Agrarian Essays

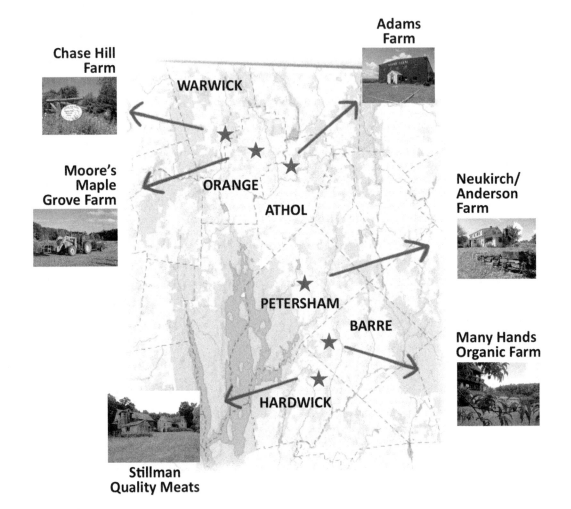

contents

six Quabbin farms vi
generous farmers deserve our business: a foreword by Leigh Youngblood ix
valuing Quabbin farms: an introduction by Cathy Stanton 1
young farmers with an old problem: the Neukirch-Anderson farm, Petersham 7
at the center of the regional food system: Adams Farm, Athol 12
farming at the edge of industry: Moore's Maple Grove Farm, Orange 18
narrowing the gap: Stillman Quality Meats (and turkey farm), Hardwick 24
back to the land again: Many Hands Organic Farm, Barre 31
de-industrial dairy: Chase Hill Farm, Warwick 37
cooperation and community: the future of Quabbin local food 45
a timeline for Quabbin area farming history 52
Quabbin farm and coop recipes 57
a note on sources 63
acknowledgments 67
about the author 69
about the photographs 71
photo credits 73
colophon 77

generous farmers deserve our business

a foreword by Leigh Youngblood, executive director
Mount Grace Land Conservation Trust

I've recognized first-hand how generous with their sweat farmers of the Quabbin area have been with the rest of us. Usually this truth is seen in the eyes and hands and heard in the voice, not the words, of the typically modest or, you could say, pragmatic farmers hereabouts.

As we wind around the hilly terrain that characterizes towns here, all too often we take for granted the fields, stone walls, and buffering woods that line most of the small highways, rough paved roads, and inevitable unpaved lanes. How right it feels to repay farmers now through the increasingly popular direct purchase of local produce, meat, and locally grown and manufactured products like jams, salsas, and frozen soups, dinners, and more, cooked and sold at the Diemand Farm Store.

With learned skill and plenty of heart for her research, Cathy Stanton tells us the backstory to a sampling of farms in the center of Massachusetts. In doing so, she brings forth the flavor of the place we call home. I thought having worked with farmers and other landowners in the greater North Quabbin area for two dozen years, I was familiar with most of the people and communities listed in the table of contents. With a mix of historical facts and personal anecdotes, Cathy has added intriguing dimensions that fill out my own experiences much more fully. My guess is that even the families profiled here will discover dimensions of context and detail they had not known or considered about their own family and farm histories.

We call her Cathy and not Dr. Stanton, PhD. In these pages, we are fed not just by recipes, many that are both healthy and economical, but also by Cathy's anthropologist's curiosity and academic rigor. It's been fun to cross paths with Cathy as she has researched, teamed up with, written about, and portrayed specific people and places I know from the perspective of permanent land conservation. In our neck of the woods, that word doesn't usually mean preservation. Most often, conservation land here is meant to continue to be available for rural living: for growing food and fuel, for exploring, and for revisiting. Revisiting the idea of a place known in the past that has been lost to modernity. I hear that a lot.

The subjects in this book speak of interdependence and mutual benefits as our way of economic and social life. Literally, farmers see the success of a neighboring farm as more of an overall benefit rather than a point of competition. Looking back over my own years here, I see such principles in the tapestry of community Mount Grace Land Conservation Trust has played a role in weaving, including Seeds of Solidarity Farm and the North Quabbin Garlic and Arts Festival, Johnson's Farm Restaurant, the Farm School, and Red Apple, Moore's Maple Grove, Chase Hill, Sweetwater, and Diemand farms, as well as Bea Riley's Homestead, and Stillman Quality Meats at Stanley Bartoszek's old place, and the Quabbin Harvest food coop, among many others. Thanks to Cathy Stanton for enhancing this fabric with stories that connect national trends in prior eras to our sparsely populated and hardscrabble part of the state. I believe others, too, who know and love this place already, will find their appreciation for its integrity, ingenuity, and continuity deepened.

valuing Quabbin farms

an introduction by Cathy Stanton

Food brings things—and people—together. Food is something we all participate in, not just because we need it in order to live, but because it's often pleasurable, interesting, and social. It can give us a starting point for making many kinds of changes that we might want to see in our world.

For me, food has been a starting point for understanding how the past and present connect in the Quabbin region. It has helped me think about how we use energy and how our economy has changed—and struggled—over time. And it has helped me evolve as a writer and educator by linking my work as a historian and anthropologist with community projects close to my heart.

I'd like to start by tracing some of those connections and how they led to the book you hold in your hand.

About ten years ago, I helped launch a community organization called North Quabbin Energy. At a moment when growing numbers of people around the world were starting to talk seriously about problems associated with a fossil-fuel-dependent way of life, we wanted to educate ourselves and others about issues relating to energy use—and not only to educate, but to support real changes in our own behaviors and others'.

The problem is that once you start thinking about all the things that fossil fuels connect to—transportation and work patterns, agriculture and food, heat and light, policy and politics—the challenge of making real change can feel daunting. We found ourselves spending a lot of time urging people to do things like turn off the lights when they left a room or carpool rather than driving alone.

These are good things that we all know we should do. But they make life less convenient, and they tend to make us feel guilty when we *don't* do them. Guilt isn't a great motivator, as any dentist who has tried to get patients to floss more regularly will tell you.

But then there was food. Making energy-related changes around food—for example, preserving seasonal local food rather than buying out-of-season produce trucked or flown from thousands of miles away—felt positive and healthy as well as productive. So North Quabbin Energy embraced food as a crucial way to talk about energy.

We began producing an annual local food brochure for the Quabbin region. We held fundraisers by selling and sharing recipes for "no-bake" cookies. At the annual North Quabbin Garlic and Arts Festival, where we organize a roster of renewable energy presentations, we feature speakers on gardening, farming, and food preservation—topics that have proved among the most popular with festival goers year after year. Many of us have made changes in our own lives by growing more of our own food and doing more to support local businesses that work to shorten food chains and invest in the local food economy.

All of this was happening as I was making a professional transition from free-lance writing and teaching adult education to becoming an anthropologist, public historian, and college professor. Over the nearly thirty years I've lived in the Quabbin area, I've been more and more drawn to a set of questions about how people use the past to understand the present and envision the future—questions I've always found intriguing and that I've explored in tandem with my deepening familiarity with New England and this part of Massachusetts. Becoming a cultural anthropologist—someone who's interested in how humans make meanings and how those meanings shape the collective ways we behave—has let me move that fascination into a new livelihood as well as a new way to contribute to the community I'm part of.

About five or six years ago, all of this started to converge. The more I thought about how we understand the past of farming in this region, the more I realized we *don't* really understand it very well. We have a set of inherited ideas about agriculture in New England: the soil was poor, farmers moved west, people turned to industry instead. But that broad outline doesn't begin to capture the struggles, adaptiveness, and sheer

Crops at Petersham's Sweetwater Farm include flowers.

persistence of New England farmers over the two hundred years since commercial markets began to dominate our food system.

Present-day efforts to reshape and re-localize farming systems can seem strangely distant from that complex history. Those efforts, often referred to as the food movement, include a wide range of things that sometimes overlap and sometimes clash—upscale eateries and farmworker labor organizing, hundred-mile diets and fair trade coffee from halfway around the globe. But at bottom, most people involved in those projects are working to rethink the industrialized, energy-intensive, heavily subsidized, and corporate food system we have now. People within the food movement try to move toward something smaller scaled, more transparent, healthier, and fairer for everyone in the system and for the natural resources that we depend on.

Wendell's Diemand Farm looks out over the landscape.

These efforts are exciting. But they're very much more focused on the future than on the past. That's understandable. But I saw a missed opportunity for exploring more fully how we ended up with the system we're trying to change—not to mention finding out what happened in earlier eras when farmers and others wrestled with these same questions.

And so, for the past few years, I've been learning more about how people have practiced agriculture in central and north-central Massachusetts over the past two hundred years. I've been piecing together stories about specific towns, farms, and people and trying to understand them within the long progression toward—and sometimes away from—large-scale industrial agriculture.

It's been a challenge to see through the many layers disconnecting us from these histories. These days, most of us have little or no direct experience of farming. And those inherited ideas about New England being unsuited to farming get in the way of seeing the realities of how people have actually farmed here over time.

To put together a clearer picture, I've drawn on a wide range of very small details and linked them with other people's research and scholarship (see "A Note on Sources" at the end). I've written the stories in this book in a style that I hope will appeal to a general reader with an interest in the subject, but they also rest on a solid foundation of scholarly exploration.

I started this project with my own town of Wendell (you can read some of the stories from that research at landcestorproject.org) and expanded in the summer of 2015 in a project called Farm Values: Civic Agriculture at the Crossroads, undertaken in partnership with Mount Grace Land Conservation Trust and funded by a grant from Mass Humanities (you can learn more about the Farm Values project at farmvalues.net.) Mount Grace was then conducting a pilot survey of food system resources in six towns at the center of its twenty-three-town service area. Farm Values focused on one farm in each of the six towns—Warwick, Orange, Athol, Petersham, Barre, and Hardwick—as a way to connect the present-day survey with a sense of the challenges that farmers have faced over time in this part of Massachusetts.

We should not underestimate those challenges. Our resources in land and wealth have always been more limited than in other parts of the state. We're not close to major cities and navigable waterways. We have a population spread out across many small towns and a good deal of hilly land not suited to growing market crops on a large scale. We're also engaged in an ongoing struggle for economic recovery after the loss of major manufacturing in many of our towns.

But the histories of these six farms show us how people have coped and changed and continued to find ways to stay on the land—or to return to it as new and experienced farmers have done again and again. These local stories of food and farming also show us what it means to work within finite resources and what it takes to create new opportunities at a small but sustainable scale that's rooted in community. The stories remind us that both food and energy ultimately rely on a natural world with considerable limits, something we often manage to forget in an era when fossil fuels have let us live well beyond our ecological means.

I hope the stories and photographs in this book will inspire you to become better acquainted with not only the six featured farms but the many others around the Quabbin that are producing good local foods and maintaining farmland in agricul-

At Seeds of Solidarity Farm in Orange, Ricky Baruc bags greens grown using intensive, no-till methods.

tural production. The recipes in the book—many of them from the featured farmers themselves—are offered as another way to connect with what's happening around our local food economy.

Most of all, I hope the book will add something to our collective conversation about what it means to eat and live more locally and sustainably—a conversation that matters deeply to us all.

young farmers with an old problem: the Neukirch-Anderson farm, Petersham

On a cold Sunday morning in February 1787, several thousand soldiers at the end of a brutal march through a blizzard surprised a sleeping army encamped in the center of Petersham and routed it quickly, all but ending an armed uprising by farmers in the central and western parts of the state. Just to the south was the farm of Benjamin Chandler, who had come to Petersham from Westford. We don't know exactly when Chandler moved to town or where he stood on the politics of the Regulation, as Shays' Rebellion was known to many. But we do know something of the subsequent history of his farm, including its 2014 purchase by young farmers who were able to leverage support from emerging models of land stewardship designed to keep farmland in production.

The two events are distant in time, but a thread connects the eighteenth-century fight with the twenty-first-century sale. Both respond to a fundamental problem in American farming: having a farm doesn't necessarily equate to having enough capital for a farmer to stay afloat in a market-oriented economy.

First, some of the big picture. In this part of the US, land has been in limited supply for a long time. As far back as the colonial period, there wasn't enough prime farmland in Massachusetts for everyone who wanted to farm it. By the time of the Revolution, many farmers—like Chandler—moved into less fertile areas like the hill towns of central and western Massachusetts.

The value of land has, of course, risen and fallen with economic ups and downs, and there have been times—for example, during the Great Depression of the 1930s—when new farmers could find fertile and affordable land in central Massachusetts. But by then, the region's agricultural economy had been struggling for many decades with the fact that food was also being produced on much larger farms elsewhere. As a commodity, industrially produced food has gotten steadily cheaper to the point that Americans spend proportionately less than anyone else in the world on what we eat (around ten percent of disposable income, on average; "America's Shrinking Grocery Bill," a 2013 infographic from Bloomberg.com, shows this drop over the past thirty years). It's simply not possible for a local farmer to compete directly with supermarkets that can sell food at low costs made possible by gigantic economies of scale.

Area farmers have historically found ways to adapt and cope. They've shifted to dairy and fruit production, formed cooperative organizations, courted tourists, and sold some of their land for residential development. But as the twentieth century went on and especially after the rise of interstate highways, long-distance refrigerated trucking, and supermarket chains after World War II, small farmers in this part of the world found themselves stuck in a financial squeeze far from resolved today. Without the benefit of such giant economies of scale, individual small farmers can't hope to compete on price. A small farmer paying a hefty mortgage has even less chance of breaking even, let alone making any kind of profit.

It is, in fact, the very same dilemma that Daniel Shays and his Regulators attempted to regulate: the balance between the demands of a capital-oriented economy and the realities of making a living from the land. So to bring things back down to the local scale: does the recent purchase of the old Chandler Farm in Petersham by Emily Anderson and Tyson Neukirch suggest a new approach to the long-standing predicament?

Emily and Tyson are part of a rising generation of young farmers who see themselves contributing to a deep rethinking of how and why we farm. Emily is a Petersham native; Tyson grew up on a conventional farm in Nebraska, learned organic farming in Oregon and California, and now teaches others to farm at the Farm School in Athol. Living on a two-acre property, they wanted to broaden their agricultural presence in the town while remaining close to its center. That's a tall order in Petersham, where large parcels seldom come up for sale and are far from affordable when they do.

YOUNG FARMERS WITH AN OLD PROBLEM: NEUKIRCH-ANDERSON FARM IN PETERSHAM

Tyson Neukirch surveys a field at his Petersham farm.

But the former Chandler Farm, which passed through ownership from Chandler descendants and others in the LePoer and King families by the mid-twentieth century, presented a unique opportunity—if the couple could put together the capital to buy the land at a price that wouldn't prevent them from ever making a living from it. Gil and Linda King owned the land and kept its central open area—the only prime agricultural land in the 106-acre parcel—as a hayfield. They had not otherwise cultivated it. The Kings had already put the farm on the open market but readily took it off again when Emily and Tyson approached them with a proposal for a different kind of sale.

Their proposal was based on another kind of big picture, one that considers the resources of a town or a region more holistically. Emily and Tyson's approach asks how local resources can fit together to support a wider range of values involving not only human sustenance and livelihood but also species and habitat biodiversity, carbon sequestration, water and air quality, and sense of place shared and appreciated by residents and visitors. This list of benefits brought together public and private entities to facilitate the sale and protect the land.

Seeing a chance to provide access to town-owned conservation land, among other benefits, Petersham's Conservation Commission quickly updated its open space and recreation plan. They saw the plan as a step toward applying for grants through the state's Executive Office of Energy and Environmental Affairs and the Quabbin-to-Cardigan Partnership, a regional landscape conservation effort with Petersham as one of its southern end points. The East Quabbin Land Trust assisted with grant applications and private fundraising, demonstrating the role that land trusts are increasingly playing in filling gaps in what individuals and public agencies can do. And the Kings themselves, along with local Realtor Chuck Berube, were key partners, willing to forego a more straightforward traditional sale in order to work with buyers who had a compatible vision of the farm's place within its social and environmental ecosystems.

Unlike the Chase Hill Farm conservation process described in a later chapter, the Petersham transaction was not a matter of expanding an existing farm operation but of establishing a new one with a young farm family. The average age of American farmers continues to rise while their numbers continue to dwindle. Despite the much-heralded turn toward local and small-scale agriculture over the past decade, the most recent federal Agricultural Census, completed in 2012, shows a considerable drop in the number of new farmers in the previous five years. While opinions differ on whether the change indicates a crisis in the making, it raises important questions. Many advocacy groups have identified access to land as one of the main obstacles facing young and entry-level farmers.

Tyson's upbringing on a conventional corn and soybean farm has given him a different perspective from those who tend to demonize the industrial food system. He has a clear understanding of how large-scale farmers, too, are caught in a system where operating costs and food prices are too often out of sync. Noting that tax codes and subsidy systems practically dictate that farmers stay on the treadmill of expansion and high-input methods, he says, "Those are not ideas conceptualized by an agrarian society. Those

Tyson Neukirch

The vision for the future of this farm goes further, upending the usual human-centered focus and instead centering on the capacities and limitations of the land itself. "This is all an experiment," says Tyson. "The exciting thing is that I think as we shift our understanding of what agriculture is and what a farm really is, we also begin to understand that the products that a farm is able to produce are going to change and evolve over time. Hopefully that's not based just solely on what the consumer demands are, because that's how we're driven right now, but what the ecological system needs of the farm itself are."

Tyson's vision nods toward the small scale of agriculture necessarily practiced by the Regulators of 1787. But Tyson is far from romantic about those older modes of farming, which he describes as "by no means restorative or resilient." Noting the inherent limitations of the Neukirch-Anderson Farm with its "little island" of prime soil in the midst of a rocky and forested terrain, he asks, "How can we work with these types of landscapes to make them both ecologically resilient and also economically productive? I think it's going to take some different perceptions of what agriculture is to be able to make that a reality."

are the ways an agricultural system interacts with the larger economic one." We're not going to get off that treadmill, he insists, until we can find a way to separate the value of land as a speculation from its value for producing food.

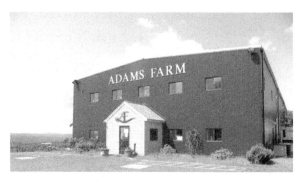

at the center of the regional food system: Adams Farm, Athol

There were five local slaughterhouses in Athol when Beverly and Lewis Adams began to sell packaged meat as part of a transition away from the increasingly unprofitable business of dairying. Lewis dispatched the cows and pigs they raised on their Bearsden Road farm, an Italian-American butcher from Donelan's market in Orange did the cutting, and Beverly packaged the meat for her husband to sell around town. The business grew, and after Lewis died suddenly in 1973, Beverly kept it going as a way to support her five children.

She also continued a trust from her mother-in-law, Hester (Comerford) Adams. Hester was born in Athol in 1885 and grew up on a farm a little farther along Bearsden Road from the spectacular spot at the crest of the hill occupied by the present-day Adams Farm. Hester's father, Richard Comerford, was the son of Irish immigrants. Richard became a veterinarian in an era when horses provided much of the transportation in both urban and rural places. Injured city horses were often sent to the country for rehabilitation, and Richard seems to have been among several animal doctors in Athol—his son became another—who attended to them.

The Comerford women equally involved themselves with both doctoring and animals, a facet of this farm family that continues into the present day. Hester was the youngest of nine children; her oldest sister Ethel pursued a medical degree in the early twentieth century, interning at Massachusetts Homeopathic Hospital in Boston and later

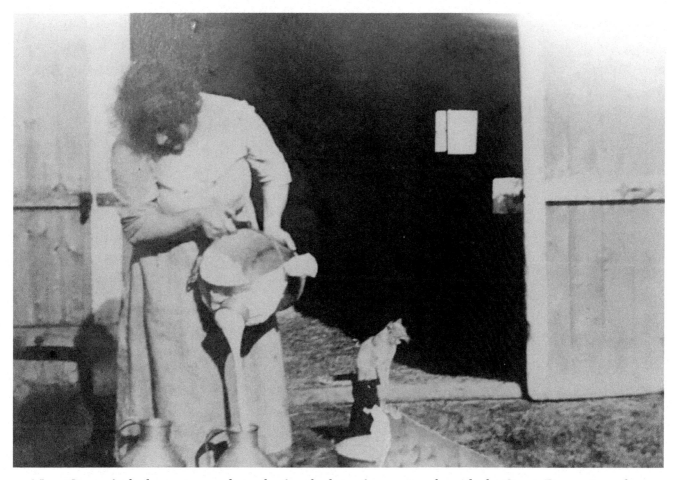

Nora Comerford, the mainstay, kept the family dairy farm going through the Great Depression when census records list her as a milk dealer, farmer, or farmerette.

running her own practice in Leominster and on the North Shore. Homeopathy, now an alternative medical practice, was much more mainstream at the time, and many women were homeopathic MDs.

And Nora, two years older than Hester, became a farmer in her own right, working with her brother Tom to keep the Comerford dairy farm going through the challenging 1920s and 1930s. Like many other area farm families, Nora and Tom sold to people in the industrial towns and sometimes worked there themselves—in Tom's case, in one of the local tool factories. Nora was the farm's mainstay: Athol street directories and federal census records list her variously as a milk dealer,

It is not known which of these sisters is Hester (Comerford) Adams and which is her older sister, Nora, but the battered family photo inspires three generations of women who run Adams Farm in the twenty-first century.

a farmer, and a farmerette, a term used during World War I for young women who took over farmhand jobs once held by men who were away at war.

In 1915, at the age of thirty, Hester married a man from a French-Canadian family, Lewis Adams (originally the more French-Canadian Louis). Hester, not her husband, seems to have remained the primary farmer and landowner. Four years after her marriage, she purchased some land next to her family's farm and used it to pasture and milk her own dairy cows. The couple had three sons—Hector, Charles (Chuck), and Lewis, Jr—but the marriage did not last. By the end of the Depression and continuing into the 1950s, it was Hester and her two oldest sons working the farm. The sons also inherited the neighboring Comerford farm when Nora died of breast cancer in the 1940s.

Hester's youngest son, Lewis, built the first slaughterhouse at the top of Bearsden Road. That was in 1946, at a time when it was already challenging to keep a small-scale dairy profitable. When Lewis married Beverly Heath in 1955, it had become clearer that they would have to adapt and change for their business to survive. Beverly has a clear recollection of her new mother-in-law entreating her to keep the farm in the family. It was a charge that struck more deeply when Hester died of a sudden heart attack that year at the age of seventy.

Lewis and Beverly worked together to reinvent their hilltop operation. Lewis had always raised some heifers for meat, and they expanded the beef herd, added pigs, and began growing corn to feed the animals. They built a new slaughterhouse in the late 1950s and stopped dairying altogether. They found customers among farmers needing slaughtering services and—once they added a small retail store—shoppers looking for good affordable meat. In the 1960s, they sold the old Comerford farm to reduce their tax burden and concentrated on raising their family and expanding their new specialized trade.

In the early 1970s, they had just updated the slaughterhouse again, building a larger facility on the eastern side of Bearsden Road, when Lewis died suddenly. He left Beverly with five children under seventeen and a challenging business to run. She kept it going, eventually taking her children into partnership with her, and Adams Farm remains in family hands, with some of Beverly's grandchildren now involved as well. Beverly still comes to work most days, although her daughter Noreen and son Rick now run most of the day-to-day operations.

They do so in a greatly-enlarged building, one completely rebuilt after a devastating fire shortly before Christmas in 2006. The fire burned the 1970s structure to the ground, but provided an opportunity for a further redesign. With more consumers paying attention to the way their food is produced, farmers increasingly looked for a guarantee of humane treatment at processing facilities. Adams Farm worked with noted livestock consultant Temple Grandin on the design of the new slaughterhouse, which is laid out in a way that works to keep animals as calm as possible as they move from trucks to the killing room.

The rebirth of Adams Farm is a source of pride for many people in a town that still struggles to define itself after the loss of much of its major manufacturing in recent decades. In fact, in 2016 the slaughterhouse was the largest in New England. It is also a vitally important node in the local food economy not just for Athol and the Quabbin area, but for the entire region. More than five hundred small farms from around Massachusetts and the surrounding states avail themselves of its services. Its retail store is almost a mini-supermarket in its own right, although

The remarkable view from Adams Farm takes in a solar field installed soon after the slaughterhouse underwent its extensive, unanticipated update after a 2006 fire. The solar field provides energy to the grid. Adams Farm also has its own solar panels.

one with a very heavy emphasis on meat.

Many farms that send their animals to Adams Farm go to great pains to emphasize the beauty of their own pastoral settings, the health and individuality of their livestock, and the aesthetic pleasures of the meat they sell. Adams does none of this: it is a resolutely utilitarian business, practical in an unpretentious way. It is, in fact, a paradoxical place, both local and not local, agricultural but also in some ways industrial, a functional animal-killing operation in a stunningly beautiful location. It is crucial to today's rapidly expanding and often upscale food movement in New England. It is, however, itself firmly rooted in an older agricultural economy and tied to the blue-collar identity of the area.

Few of its retail products are local: much of the beef in the store comes from Amish farms in Pennsylvania, while most of the pork is from Canada.

Noreen Heath-Paniagua, Beverly (Adams) Mundell, and Chelsea Frost, from left, carry on the Adams Farm legacy with foresight and dedication.

Rick Adams makes long trips each week in the Adams Farm truck to pick it up. But all of it is hormone and antibiotic free and processed on-site, placing Adams Farm somewhere between old and new, organic and conventional, artisanal and industrial.

It might seem that the four-generation legacy of women centrally involved in running Adams Farm and its slaughtering operation is somewhat paradoxical as well. But the women themselves don't see it that way. For Beverly (now Mundell), the business is what was in front of her when she needed to provide for her family. For her daughter Noreen Heath-Paniagua and Noreen's daughter Chelsea Frost, this is simply what they've done all their lives. "It's all I know," in Noreen's words.

They do admit to finding inspiration in a photo of Nora and Hester Comerford standing next to each other looking as though they've just finished milking. The photo hangs in several places around the building, one of only a few family photos that survived the 2006 fire. "They're like my heroes, just seeing how hard they work," Noreen says. "That's the motivation that some of us need to keep going sometimes."

farming at the edge of industry: Moore's Maple Grove Farm, Orange

John and Laura Moore met fifty years ago at a 4-H Club gathering in Washington, DC. Both were teenagers from farm families—hers in Michigan, his in Orange, Massachusetts— and they were sent to Washington in recognition of their prize-winning farm products. They fell in love, carried on a long-distance romance, married, and settled down two miles north of the center of Orange on the Cross Road farm where John was born. They have lived and farmed there ever since, raising five children who still live close by. Their grandchildren are the ninth generation of Moores to live in this part of town.

It is as rural and pastoral a story as can be imagined. It seems—and the farm feels—very far removed from the center of Orange in either its industrial heyday or its struggling present. And yet Moore's Maple Grove Farm has reflected the changes not only in Orange but in the area's larger industrial economy for well over a century. Linkages between them challenge us to see the farm and the town as two sides of the same story.

That story began for the Moores not long after the American Revolution on another farm a little farther north on Jones Cemetery Road. The family formed part of a tight-knit network of North Orange farm families for generations. Like their neighbors, they produced things mainly for their own use but also some for sale in commercial markets: meat, wood, hay, wool, butter. Those markets would mostly have been local, and for small farmers like the Moores, quickly expanding mill towns along the area's river valleys provided new venues for their products.

Factories intersected with the older farm economy in other ways. Good farmland had been hard to come by for many decades, and not all the sons and daughters in farm families could move onto farms of their own. Some went west, of course, in the fabled movement toward the frontier. Many moved to the rapidly growing cities of New England and beyond. But many stayed closer to home, finding work in the mills but not necessarily abandoning their old homes or networks of connection. In the Quabbin area, the mill towns never sprawled into giants

In the 1940s and 1950s, John Moore II employed draft horses and a sledge for gathering maple sap at Moore's Maple Grove Farm.

and the older farm world was never too far away. Census records show many examples of people who moved from one sector to another and sometimes back again, piecing together lives and livelihoods that combined the agricultural and the industrial.

One such family, the Marbles, came from Northampton around 1895. Alden, Judson, and Joel Marble grew up on a farm but seem to have been among those who took non-farming jobs as young men. For Joel, the result was tragic: he died in an accident at the sawmill where he worked, leaving a young wife behind. Alden's story was equally sad: he married in his mid twenties and was the father of a three-year-old girl, Mary Charlotte (known as Lottie), when his wife died suddenly of meningitis. Along with their then-widowed mother, this clan of bereaved Northamptonites moved to Orange and Athol together, perhaps seeking a fresh start. The men in the family worked at the New Home Sewing Machine Company in Orange, and Alden remarried when his daughter was fourteen.

But the Marbles, especially Alden, seem to have been drawn back to farming. By the turn of the century, he and at least one of his brothers had bought the property on Cross Road from an older Orange family, the Goddards. They paid off the mortgage very quickly—an advantage, perhaps, of being able to earn year-round money in the factories.

The Marble brothers didn't keep the farm long. Lottie Marble was by then being courted by the young John N. Moore from the Jones Cemetery Road family, and the young couple eventually took it over. Their grandson John Moore III tells the story of how the eager suitor walked to see his sweetheart and how the sight of a stand of sugar maples at the edge of the Cross Road farm was his sign that he was almost there. John surmises that may have been the reason they renamed it Maple Grove after they married and settled there in 1903.

Having gone back to the land, Alden Marble continued to farm until his death in 1935. He bought a smaller piece of land between Orange and Athol and may have made a living as a market gardener by selling directly to customers or to some of the many small grocery stores that serviced the two towns by then. John and Lottie Moore ran the Cross Road farm for many years as Maple Grove Dairy with a substantial sideline in maple syrup. They were one of more than thirty small dairies in Orange that delivered milk to customers' homes well into the middle of the twentieth century.

By then, John's father, John II, had taken over running the farm with his wife, Jeannette, and eventually the help of their son John III. Like others in the area, they found as the decades went on that selling liquid milk at a small scale was no longer economically feasible with the advent of bulk tank collection, new regulatory hurdles, and increased competition from larger companies. After John III took over active management of the farm in 1999, it didn't take long to decide that they should make a shift toward more mixed production—the kind, in fact, that both his Moore and Marble forebears would have been very familiar with.

In 2016, John grew vegetables, raised beef cattle, and sold hay and maple syrup. Until 2015, his wife Laura ran a small bakery, Maple Grove Farmhouse Bakery, in the former Minute Tapioca Factory. Since closing her own bakery, she works out of a church's commercial kitchen and sells baked goods at farmers markets. Their children work in education and social service, not agriculture, but they help out when needed. Their twelve-year-old granddaughter Sienna has run her own flower business for several years. Essentially, though, John and Laura point out, their farm and bakery are both one-person operations with aging proprietors. It's not clear whether the long history of Moores farming in North Orange will continue beyond the present generation.

The Moores' continued presence in local food and farming shows the kind of opportunism and adaptability that small farmers have always needed in order to stay in business in this part

Laura Moore's bread and bakery business augments the farm's income.

Sienna Moore helps her grandfather with haying.

of Massachusetts. For example, John, who learned to drive draft horses when he was just a boy collecting maple sap with his father and grandfather, did not stop using horses when the family added tractors.

As their story shows, members of this family have long seen the industrial towns as both markets and sources of off-farm work. Even the supermarkets that have in many ways made it so difficult for small farmers to compete have served a useful function for the Moores. John's mother worked in some of the local supermarkets, and although Laura Moore learned to bake in her mother's kitchen, she honed her business and production skills during years of working at supermarkets in Orange and Greenfield.

Their business draws deeply on long-established relationships with neighbors and family but also on newer and revitalized networks and ideas. All three generations join together to sell at farmers markets in Athol and Orange. The Moores' meat and vegetables can be found at the Quabbin Harvest Coop in Orange. For this longtime farm family, the emergence of local food enthusiasm has dovetailed neatly with their move toward more diverse offerings. They have also embraced farmland preservation strategies, protecting their more-than-four-hundred acres with the help of Mount Grace Land Conservation Trust. Whatever the younger generations of Moores decide to do, the land will remain in agricultural production.

John Moore carries on a sturdy tradition.

As the area's mill towns redefine themselves, farms like the Moores' may have an important role to play, knitting agriculture more fully back into a story dominated for many decades by empty factory buildings and persistent economic struggles. Seeing farms and factories as two sides of the same coin may open the way to a vision of a more hybrid and locally rooted economy that draws on its people's strengths and the land they share.

narrowing the gap:
Stillman Quality Meats (and turkey farm), Hardwick

On a humid July day, the barn at the first of Kate Stillman's two farmsteads in Hardwick is filled with cheeping. Dozens of young turkeys with pink, perplexed faces occupy one side of the nineteenth-century structure. Their short lives will end a few months later at a second farm in the southern part of town, where Kate has recently added a poultry abattoir that gives her greater control over the growing meat business she started ten years ago.

The presence of turkeys links Stillman Quality Meats with the past of this farm at the corner of Thresher and Jackson roads. It also represents a revealing moment in the longer history of agriculture in central Massachusetts. Area farmers found it harder to compete

Young turkeys roost comfortably at Stillman's.

with large-scale industrial food production, especially after World War II. Many owners of small farms shifted from full-time to part-time or seasonal operations and sometimes to

In 1888, the Mahan family took their places for a formal photograph.

hobby farming that removed them from commercial markets altogether. But when farmland has been owned by people who have cared about keeping land *as* farmland, the properties have remained as potential resources for newer enterprises, like Kate's, that are part of today's attempt to rebuild more locally or regionally oriented food systems.

In the 1820s, what seems to have been a modest eighteenth-century farm came into the hands of the Cleveland family, who owned a good deal of property in the northern part of Hardwick. In 1869, they sold it—by then expanded to a 160-acre parcel—to Michael and Ellen Mahan, a recently married young Irish couple.

The Mahans appear to have been highly skilled and energetic farmers. They accumulated the capital needed to buy and run a farm within ten years of immigrating to the US. In their first year at Thresher Road, with minimal equipment and only a twelve-year-old hired boy to help them, they were producing corn, wheat, and potatoes, all common cash crops at the time. They also made an impressive three thousand pounds of cheese, suggesting that Ellen, who at thirty had just delivered the couple's first child, was, like many farm women, an adept cheesemaker. They also sold meat and wood. Ten years later, the 1880 census showed that their "unimproved" acres had been converted to income-producing pasture, orchard, or woodlot and that they had three sons and a daughter.

The Mahans' mixed farming operation resembled countless others in central Massachusetts in that time period. Profits from meat and grain dwindled when railroads began bringing products from much larger farms to the west. But people in Hardwick found new sources of income not only from the textile and paper mills that appeared by mid century but by a significant level of both on- and off-farm cheesemaking. Three cheese factories—small-scale operations that bought milk from a number of nearby farms—operated in Hardwick by the 1850s and accounted for more than half of the three hundred thousand pounds of cheese made annually in the town.

By the time Michael and Ellen Mahan set up their farm, cheesemaking, too, felt competitive pressures from western and midwestern states. We have no detailed data about what the Mahans produced in the later years of the nineteenth century. But if they continued to follow the general local pattern, they likely shifted more toward liquid milk dairying, perhaps with an increased emphasis on their orchards, poultry, and market gardening.

Michael and Ellen died within a few years of each other in the first decade of the twentieth century. Their son Daniel and daughter Mary kept the farm going for several more decades. Mary died in 1932, and the 1940 federal census shows Daniel living there with a hired farmhand. When Daniel died the following year, the youngest Mahan son, John, then in his sixties, began returning to Hardwick from his home in Springfield where he had long been a traveling salesman in the food industry. John was the only one of his siblings to marry. A 1920s family portrait of him with his family gives a clear sense of the middle-class aspirations of the one-time smalltown farm boy. And yet he seems to have been unwilling to let the old family farm go.

John had an ally in his efforts to keep farming: his son-in-law, George Crombie. George was an Irish-American from Connecticut whose family owned an ice delivery business and also operated a fleet of school buses. Although George attended

George Crombie, John Mahan's son-in-law, raised award-winning heifers as a 4-H member in the 1930s.

college and worked in the family companies, he himself seems to have wanted nothing more than to be a farmer. As a boy, he belonged to 4-H, raised heifers, and won an award from the state for his agricultural accomplishments. When his wife's old family farm needed stewardship, he was immediately on board. He lent a hand with haying and bought his father-in-law "a couple of turkeys to keep him busy," in the words of George's daughter Susan Twarog.

Before long, the couple of turkeys had expanded to a seasonal business that saw the two men raising as many as five thousand free-range birds annually. They bought day-old chicks in Connecticut and kept them inside the barn until the Fourth of July, then put them outside with fencing and flares to keep the foxes at bay until the turkeys were big enough to protect themselves. After John Mahan died in 1956, George kept the business going, slaughtering the birds in the

John Mahan, left, and George Crombie (along with a young helper) transport a load of hay in 1942.

barn until about 1960 when more stringent regulations required him to build a separate government-inspected processing plant on the property.

In his work for regional food wholesalers, John Mahan had helped widen the gap between small-scale producers and those who were consolidating food production and distribution. He and his son-in-law weren't able to compete in those expanding markets. George's daughter Susan notes that her father was lucky if he broke even at the end of a year. But the two men—and a generation of young cousins who loved spending their summers at the farm—seem to have been motivated more by passion and place-attachment than by the profit motive. And they *did* find buyers for their birds: they sold turkeys to stores, restaurants, hotels, and hospitals and did direct-marketing to customers as well. It was laborious and old-fashioned in its reliance on hand labor and hand selling rather than mechanization and wholesaling. But it kept the farm functioning.

There has been only one gap in this farm's operations: between 1975, when George and Elizabeth Crombie finally sold the property out of the family, and 1985, when it was bought by Bob and Linda Paquet. The Paquets weren't commercial farmers, but they raised sheep for wool, restored and renovated the house and barn, and saw themselves as carrying on the Turkey Farm name and presence in town. Their children belonged to 4-H. And the family entered a three-tiered float in Hardwick's 250th anniversary parade that presented a living rendering of the old farm's story. The first layer showed a display about the Mahans' dairy farm. A live turkey on the second layer represented the Turkey Farm years. The Paquets' pet ram, Napoleon, rode on the top.

With its current owner, who bought it from the Paquets in 2005, the story takes another turn. Kate Stillman's grandfather was a dairy farmer in Lunenberg, Massachusetts. Her father Glenn started out there, building an extensive direct-marketing vegetable business that now encompasses farmland in Lunenberg and Hardwick's neighbor New Braintree.

The Stillmans coordinate their efforts and collective presence in markets around eastern and central Massachusetts. Kate's brother owns a farm just north of Hardwick Center, and Kate purchased a second Hardwick farmstead seven years ago, where she built her abattoir. Year-round, the family is able to offer a wider range of products than most small direct-marketing operations. They are somewhere between small and large scales, serving customers who value a sense of face-to-face interaction but who also expect the kind of consistency and variety they're accustomed to finding at supermarkets.

"It's a hard stretch," Kate admits. "If you have people who are used to going to the grocery and fancy high-end chain stores, they just are used to seeing that produce there. And they don't understand that the peaches might not quite be ripe, you know, that the turkey didn't grow to 27.2 pounds, it only grew to 26 pounds! So being able to do our own processing actually has helped us narrow that gap a little bit and get closer to what people's expectations are."

Kate always intended to put her own name on her business. But local memory was strong enough that she found she couldn't quite shed the Turkey Farm name, and the new farm started out as Stillman's at the Turkey Farm. More recently, Kate changed to Stillman Quality Meats. The turkeys she felt obliged to add to her repertoire still fill the barn on Thresher Road in the summer and fall and find their way onto her customers' tables each holiday season. In addition to selling at numerous Boston-area farmers markets, Stillman Quality Meats also sells year-round

Kate Stillman and an assistant stock a meat case for the July 2015 opening at the Boston Public Market.

at a stall in the Boston Public Market, which opened in the summer of 2015.

Kate herself seems to have a sense of where her business fits within these complex, overlapping trajectories. "I couldn't imagine trying to do this in a different time period," she says. "It's nice being in an area that's so booming and blossoming, and it's nice to have the support."

back to the land again: Many Hands Organic Farm, Barre

Julie Rawson was raised on a western Illinois farm by parents who combined agriculture, medicine (her father was a large-animal veterinarian), and activism (her mother was deeply engaged in civil rights issues). When pregnant with her first child, she said to her husband, Jack Kittredge, "I'm really sorry, but I think I have to raise these children on a farm."

Jack's background was suburban. "I wasn't a farmer and didn't intend to be a farmer," he recalls, "but I fully supported homesteading, raising your own food, having the kids have access to all the principles and lessons that nature teaches you."

They purchased fifty-five acres of land carved out of a once-larger farm in the southern part of Barre and moved there in 1982. Surrounding acreage was already being subdivided into smaller house lots, but Julie and Jack immediately created a new farm, building an energy-efficient house and a number of outbuildings. They planted fruit trees and added three more children to their family. They started marketing food on a small level in 1984 and now sell vegetables, fruit, meat, and eggs, mostly through Community Supported Agriculture (CSA) shares. They are also deeply involved in the policy, advocacy, and education sides of farming through their work with the Northeast Organic Farming Association. Julie serves as executive director of NOFA/Mass, headquartered at their farm.

Julie Rawson, Many Hands organic farmer and executive director of NOFA/Mass, just knew she had to raise her children on a farm.

We tend to think of the 1960s and 1970s as the back-to-the-land era, making Many Hands Organic Farm something of a latecomer to that movement. And yet the history of going back to the land in the northeastern US shows that Americans have been acting on that impulse for a long time. In starting their new Barre farm in the 1980s, Julie and Jack combined many motivations that prompted others—urban, suburban, and formerly rural—to start or return to farming over the past century or more.

The first wave of new homesteading happened around the turn of the twentieth century, largely in response to repeated financial panics. American ideals of independence and achievement sat uneasily with realities of an expanding industrial capitalist economy, with its market-driven booms and busts. Buying a piece of land where you could grow food—what historian Dona Brown calls "the enduring dream of self-sufficiency in modern America"—was one way that people sought some kind of stability and security. Along with city dwellers looking for a source of food and community in hard times, recent immigrants from rural parts of southern and eastern Europe found an answer to their search for affordable land of their own in the older farms of the American Northeast.

Many such farmers were trying to reconcile nostalgia for the past with faith in a scientific future. They followed the latest developments of agricultural science as promoted by agricultural extension services. And they patronized companies selling new products to make farming more efficient and profitable: hybrid seeds, silos for year-round livestock feeding, chemical fertilizers and pesticides, tractors and other new machines driven by petroleum and electricity. In the Quabbin area, for example, the newly-married Spaulding and Carolyn Rose took up farming in 1930 at an old Phillipston farm that Spaulding's family had bought a few years earlier. There, they parlayed their college educations in pomology (the science of cultivating fruit trees) and nutrition into a commercial orchard business. Christened Red Apple Farm, it struggled through the Depression but eventually found a niche in the regional farm economy.

However, many farmers found that the costs of newer tools and methods left them with the same problem that faced Daniel Shays' rebels many years earlier: tying up all of their capital in land and equipment made it harder rather than easier to stay in business and out of the trap of debt. New England's overall agricultural production actually rose right up to the time of the First World War. But it did so because a smaller number of better-capitalized farms produced food more and more intensively while many smaller ones struggled—and often failed.

The seeds of more environmentally conscious back-to-the-land movements were sown in the 1920s and 1930s. Reformers like Sir Albert Howard and Rudolph Steiner began to argue that the solution for small farmers was not to keep chasing the cutting edge of industrial farming but to go in the opposite direction—toward smaller-scale, more energy-efficient, less mechanized and less input-hungry methods. Howard's "organic" agriculture emphasized continually feeding and enriching the soil, an approach now often termed sustainable or—more accurately—regenerative, reflecting the notion of farmland as a living organism.

By the 1960s and 1970s, such ideas were resonating with more and more people, including a counter-cultural generation connecting the dots among a host of environmental, economic, and social problems. The many back-to-the-landers of this generation believed that industrialized agriculture created at least as many problems as benefits. They also saw new (or perhaps old but rediscovered) methods of producing, selling, and sharing food as a crucial way to address a wide range of social ills. They helped create the framework for an alternative food system that was already becoming more mainstream by the time Julie Rawson and Jack Kittredge charted their own path into organic agriculture in Barre.

They did so at a time when some homesteaded farms of an earlier era were adjusting their course once again in a changing regional food economy. Some turned toward direct-marketing strategies that have since become widely favored by small producers. Farmers markets began to reappear in town and city squares, while pick-your-own operations and farm stand sales also became more common. At Red Apple Farm, Spaulding Rose had staunchly resisted the pick-your-own approach, insisting that harvesting fruit needed training and expertise. But his son Bill, who took over the orchard in the 1980s, recognized that nostalgia for rural places was as marketable a product as apples. He began the transition toward agritourism that the farm successfully pursues into the present day.

There's often an implied critique of industrial farming in such an approach. But at Many Hands Organic Farm, that critique has always been direct and political. Everything Julie and Jack do at the farm comes from their belief that the industrial food system is broken and destructive and that no single part of it can be fixed in isolation.

Like other farmers who approach food cultivation in a holistic way, Julie and Jack see the vitality of farmland itself as integral to the nutritional value of what is grown there. The concept carries over to the health of the human and non-human species who eat those foods. They believe that the worst stresses of contemporary modes of living, working, and eating fall disproportionately on poorer people, often people of color. In response, they employ and mentor ex-offenders and people in recovery from addictions. They frequently teach others about ways to preserve food at the peak of its freshness, including by canning, pickling, fermenting, drying, freezing, and root-cellaring (a sizable root cellar is one of the features of their house that they're proudest of). Over time, they've added minerals and organic matter to their farm's soil, cultivating it in ways that minimize disruption to microbial life teeming below the surface. In a time of ever-increasing concern about climate change, they've been at the leading edge of

Like others who went back to the land, Jack Kittredge and Julie Rawson combined motivations including "the enduring dream of self sufficiency."

discussions about how regenerative approaches to agriculture can help hold water and carbon in the land rather than contributing to aquifer depletion and greenhouse gas emissions.

They're also helping to ground the dreams of a new generation of homesteaders in the realities of running a farm. "There are a lot of folks out there saying, 'Oh, I just want to give up my job and be a farmer,'" Julie says. "So I say, 'Well, why don't you come out and work with me for a day and see how it feels?' And some of them go away and say, 'Okay, I get it now. I don't really want to be a farmer, but it's nice to know where my food comes from.' And others get totally taken with it and want to change their lives."

Jack heads out to the fields on a late summer day.

de-industrial dairy: Chase Hill Farm, Warwick

Chase Hill Farm in Warwick feels like one of the most peaceful places on earth. It's on one of those roads where you look twice when a car happens to drive by. There's a view of Mount Monadnock across the fields to the northeast. Brown and white cows are usually seen grazing on one of the hillsides. A mysterious little doorway—the entrance to the cheese cave, it turns out—leads into the side of a hill. The only disruption is from the barking sheepdog that comes to meet you.

The calm is partly a reflection of the farm's location. But it's also something that has been achieved gradually over the three decades since Mark Fellows and his wife Jeannette took over his parents' business. Slowly and thoughtfully, they've been disentangling what was once a conventional commercial dairy from the surprisingly complex processes that have made cow's milk one of the most fully industrialized of food products. At the same time, they're part of a broader new partnership between farmers and land conservationists, a relationship that was by no means always amicable.

The Fellowses—mother Vida, father Archie, and a growing number of children—came to Warwick sometime around 1920. Some family members had been working in factories in Orange, but they took up farming in their new home, first at Mayo Corners where Athol and Richmond Roads cross and then on Chase Hill Road just north of the current Chase Hill Farm.

In 1944, the five youngest Fellows children tended the family cows.

It was not the best of times to be a farmer in a hill town like Warwick. Although New England's agricultural output expanded all the way through the nineteenth century thanks to more intensive farming methods, approaches that worked on prime lowland farms did not suit thinner upland soils. In an effort to keep up with the pressures of expanding commercial markets, many hilltown farmers had drastically cleared and overused formerly wooded land or pastures. The widespread clearing of forested land for the area's sizable lumber industry also contributed to erosion and depletion of hilly areas, sparking a movement for land conservation and calls to let the uplands return to forest. The Commonwealth of Massachusetts bought largely denuded Mount Grace in 1915 as part of the new state forest system. After the start of the long agricultural depression that followed World War I—prelude to the Great Depression of the 1930s—more and more old Massachusetts farms became public lands and new-growth forests.

But not everyone was ready to give up on farming in Warwick. The town elected its first director of agriculture in 1919. The arrival of the Fellows family around that time suggests that some people saw the old fields as opportunities rather than problems or relics of a bygone day. Archie and Vida's large family—eventually there were fourteen children—provided an in-house labor force, and there was a living to be made in the active network of small local dairies that supplied area towns with milk.

Archie and Vida's marriage eventually failed, but Vida and the children stayed on

Vida Fellows, center back with a baby, raised her fourteen children just north of what would become Chase Hill Farm.

Oliver Fellows works a tractor in the mid 1950s.

the farm. Not only that, but her second-oldest son, Winfred, harbored dreams of starting a new farm. Like many farm children, he raised animals of his own from an early age. And like many of his siblings, he completed some of his high school training at New Salem Academy, fifteen miles away, where an agricultural studies program inspired many local youth to pursue farming. After graduating, he acquired a small herd of Guernsey cows and made plans to build a new farmstead on the hillside next to his parents' home.

World War II intervened. Twenty-year-old Winfred Fellows was drafted in 1942 and lost his life in Germany two years later. He had entrusted his herd to his next-oldest brother, Oliver, and it was Oliver, with his wife Virginia, who eventually purchased the acreage from Vida and built the house and barn that stand on the property today. They farmed it from the mid 1950s to the early 1980s, raising their three sons and selling their milk to regional bulk distributors who superseded smaller-scale, more local dairy networks of previous decades.

It was an exceptional thing for a young couple to start a new farm in a north central Massachusetts hill town in the middle of the twentieth century. It went against all the trends of old fields continuing to revert to forest or being sold for housing developments, young people moving into towns and cities for work, and food being distributed over longer distances on expanded postwar highways that let refrigerated trucks travel between source and supermarket with petroleum-powered ease. Town historian Charles Morse reported in his 1963 *Warwick, Massachusetts: Biography of a Town* that the director of agriculture "had lost his enthusiasm from lack of interest and the town refused to elect one in 1941." Yet Winfred Fellows dreamed of a new farm in the 1930s and 1940s and Oliver and Virginia realized that dream in the 1950s. Their story invites us to look carefully at the many ways farmers have found to maneuver within the trends and sometimes to push back against them.

By the early 1980s when Mark and Jeannette Fellows bought Chase Hill Farm from his parents, dairying in the area was shrinking fast. Continued consolidation in the milk distribution industry, among other factors, made it difficult or impossible to stay solvent below a certain size. Mark had been trained at Cornell University to follow a conventional path into farming. But quite quickly he and Jeannette found that they didn't want to be driven by conventional marketplace logic. Experimentally, one step at a time, they began to set the farm on a different footing.

In hindsight, the basis of their strategy has been simple: keep costs of production

Jeannette and Mark Fellows set the farm on an unconventional but successful footing.

Chase Hill cows enjoy a diet rich in green grass.

low and produce as much as possible of what they need themselves. In practice, it has meant embracing an idea that has steadily gained currency among farmers practicing regenerative and pasture-based agriculture: that healthy soil and grass are the foundation for just about everything else that happens on the farm. The forty cows at Chase Hill eat fresh grass seven months of the year and dried grass—hay—in the winter, all grown on Chase Hill's own fields plus some owned by neighbors.

"That whole thing is basically a solar panel," Mark says, pointing to a hillside whose soils earlier generations of farm agencies and educators deemed marginal. Actual solar panels on the barn supply more than half the farm's electrical needs, and three draft horses provide a good deal of the energy for spreading manure and making hay, further minimizing the need for petroleum products.

Hay alone is not nutritious enough to support a cow producing milk. The usual solution is to grow or buy corn so that cows can be milked year-round. But rather than follow prevailing practices, Chase Hill Farm went the other way. Mark began breeding and milking the cows only in the warmer months and letting them go dry in the winter. This was historically the pattern in New England before the emergence of new technologies like silos for longer-term feed storage in the early twentieth century. Year-round dairying, like industrial agriculture more generally, demands standardization and creates a steady, predictable product. But seasonal dairying is more of a dialogue between farmers' plans and the limits of what can be produced without undue stress to human and animal bodies and the land that supports them.

Two final strategies have moved Chase Hill Farm even closer to self-sufficiency. First, Mark and Jeannette stopped making wholesale milk sales that provided the bulk of their income in the early days. Instead, they moved toward direct marketing, another trend that is both new and very old. And second, Jeannette started to make cheese in 2001, adding value, variety, and year-round flexibility to the farm's other offerings of meat and raw (unpasteurized) milk. A range of artisanal cheeses comes out through the mysterious little door in the hillside, to be sold at farmers markets, some area supermarkets, and in the farm's small on-site store.

Another change in 2001 reflects a recent trend in land conservation: protecting farmland in perpetuity so that it can remain in production. When the seventy-five-acre Mayo Corner farm where Mark's father was born came up for sale, the Fellowses were concerned that one of the few remaining farms in town was ripe for development into house lots. They also saw an opportunity to gain more control over the vital pasture and hayfield base that their farm depends on. They negotiated with the state about development rights on their 260-acre farm and, with proceeds of that sale plus help from Mount Grace Land Conservation Trust, were able to buy the old farm at the crossroads.

Warwick's historian Morse was caustic about the role of the state in transforming

Mark works with the Chase Hill Farm draft horses, a vital part of the farm's low-input methods.

so much of the town's farmland into forests. The state, he declared in his 1963 history, had "gradually devoured nearly one-third of the town bite by bite" while the majority of the population "bowed to the inevitable." But protecting farmland charts a middle course between taking land out of active use and letting purely financial considerations decide its future. It's an approach that resonates with Mark and Jeannette Fellows' quiet but determined resistance to the logic of the industrialized food system. And it has enabled their ongoing work of building a thriving business at its very edges.

cooperation and community: the future of Quabbin local food

In a 2015 interview for the Farm Values project, Hardwick farmer Kate Stillman talked about her sense that small-scale farming is much stronger when farmers work together as part of an agricultural ecosystem.

"We all matter, all these little farms, we all need each other, we all survive because of each other," she told me. "So I always say when one farm's doing well it sort of starts to roll out and trickle out into the greater community."

Small farmers everywhere have long felt this sense of interdependence. On a local scale, they've joined together to hold markets and fairs to market their products and share knowledge and skills. The venerable Hardwick Fair, first held in 1762, was the first one authorized in what was then still a colony. Newer markets and annual gatherings like the regional conferences of NOFA or the North Quabbin Garlic and Arts Festival in Orange are vigorous successors to earlier events.

Regionally, farmers have formed cooperative buying and selling groups and built important pieces of agricultural and marketing infrastructure such as retail stores and shipping depots. Nationally, they've formed organizations to lobby and leverage their collective numbers. The National Grange of the Order of the Patrons of Husbandry, better known simply as the Grange, began in the middle of the nineteenth century to support and advocate for farmers. At its peak, it was a politically potent organization with chapters throughout the US, including in most of the present-day

Quabbin towns. North Orange, Petersham, and Ware still had Grange chapters in 2016.

On the other side of the food equation, eaters have sometimes banded together in buying clubs to get reliable access to the foods they wanted. In fact, coops as we know them today first appeared in England in the early industrial era when factory workers pooled their resources to buy staple foods.

For many people, the phrase food coop may evoke images of small stores with bulk bins of whole grains and organic vegetables. These ventures appeared in many places during the iconic back-to-the-land and hippie era of the 1960s and 1970s. There were several of these in the Quabbin area, including the still-thriving Leverett Village Coop. The Orange-based Our Daily Bread coop flourished from 1972 well into the following decade. Members came largely from the surrounding towns and devoted countless hours to provisioning and distributing what was then called "health food" in a succession of locations around Orange.

Credit unions are another familiar type of cooperative. Fitchburg-based Workers Credit Union occupied a building in the center of Orange for many years, but moved to a new building outside of downtown in 2014. Mount Grace Land Conservation Trust bought the old bank on North Main Street to provide a storefront for another cooperative business: Quabbin Harvest, a food coop that had been incubating since 2009 in the nearby Orange Innovation Center (formerly the Minute Tapioca factory).

Stories of the Neukirch-Anderson Farm in Petersham and Chase Hill Farm in Warwick in previous chapters show why a land trust might engage actively in the food economy. East Quabbin Land Trust has partnered with a local restaurateur to revive the Petersham Country Store as a local center for food and conviviality. Similarly, Mount Grace saw that farmland preservation was only part of the task of rebuilding a more vibrant local food system. In partnership with Quabbin Harvest Coop, it is working to create a retail hub for local food producers, especially those working on the small scale that most local farms are suited to.

Quabbin Harvest stocks as much as possible from nearby sources. In addition to selling local vegetables, fruit, meat, milk, eggs, cheese, yogurt, pickles, honey, syrup, craft and gift items, and more, the coop offers weekly fruit and vegetable shares and custom bulk ordering reminiscent of older food-buying clubs. Its storefront is open to all for year-round shopping. This central

Peter Diemand, Anne Diemand Bucci, and Faith Diemand of Diemand Farm, Wendell, from left, top, and, from clockwise to right, Steve Vieira, Angela Baglione, Lisa Kalan, Laura Davis, and Karen Davis, from left in photo, of Sweetwater Farm, Petersham; Ricky Baruc of Seeds of Solidarity Farm, Orange; and Rachel Scherer of The Little White Goat Dairy, Orange, supply Quabbin Harvest Coop in Orange and other area sellers.

marketplace means that producers can do a little less traveling and labor-intensive hand selling while buyers can find more local food in one place than is usually available at area farmers markets and farm stands.

It's an exciting venture and a fairly radical experiment for both the land trust and the coop. They're up against lower prices and greater selection and convenience of supermarket chains that weren't quite as large or dominant even in the era of hippie coops. Supermarkets have been around for a surprisingly long time—the Great Atlantic & Pacific Tea Company, or A&P, launched its first "economy" grocery store in 1912. But although they grew rapidly with around eight thousand A&P stores in the first decade and comparable growth for Piggly Wiggly, First National Stores, and other early twentieth-century chains, they didn't really reach super size or start to sell nearly everything under one roof until after World War II.

In the early decades, it wasn't uncommon to find a branch of several supermarket chains even in nearby towns. Both Athol and Orange had branches of several national chains for many years. People still shopped very close to home even though the stores themselves were increasingly headquartered in more distant places. Supermarkets were just one part of a food ecology that included independent butchers, slaughterhouses, dairies, vegetable growers, orchards, and other local producers. Well into the 1970s and 1980s, people in Quabbin area towns bought food at relatively small, family-owned stores like Donelan's, Carroll's, Stan's Sooper, and many little corner mom-and-pops.

Smaller-scaled markets often sourced some products from local farmers. But that's no longer the case with giant chains that have captured a lion's share of food sales since the 1980s. Today's supermarkets and big box stores have benefited from continued corporate consolidation in food production, processing, and marketing, along with computerized inventory systems that enabled increasingly long and complex supply chains. The big stores command immense resources that allow them to keep a supply of every imaginable food on their shelves year-round.

Small producers cannot hope to sell to gigantic companies. Although many supermarkets do now promote local food, only the largest of regional farms can supply enough to register in an inventory system geared to a global scale.

At the same time, the big stores carry things that once were available *only* in small, locally oriented storefronts and coops—organic vegetables, whole grains, health food supplements, and other alternatives that have become mainstream. Because of their vast economies of scale, big chains can offer items at lower prices than smaller retailers can, creating a paradox: locally grown produce at a locally owned store is more expensive than fruit and vegetables shipped from California, Peru, or China.

How can a tiny food coop like Quabbin Harvest stay afloat in an economically challenged area? That's an open question, and it gets to the heart of what may be possible for our local food system. As coops and farmers have always done, Quabbin Harvest counts on the power of collective action. Its experiment in locally centered marketing depends entirely on whether enough people in the area find reasons to support it. It's a difficult moment but also an exciting one.

In the same moment, other Quabbin area farms fit into local, regional, and national food systems in a wide range of ways. Shifting toward a grass-fed herd and producing artisanal cheese mostly sold in markets outside the area, Robinson Farm, a multi-generation dairy enterprise in Hardwick, has followed a path similar to that of Warwick's Chase Hill Farm. Meanwhile in Orange, the Hunt family maintains one of few remaining local dairy farms still selling liquid milk in conventional dairy markets. A visible landmark with frontage along Routes 2 and 202 and a field of solar panels covering ten acres of former woodlot, Hunt Farm

Hunt Farm in Orange carries on a generations-long tradition of dairy farming while also supporting renewable energy initiatives by hosting a solar field.

Phillipston's Red Apple Farm offers pick-your-own fruit in season and presents a well-stocked farm store.

continues to struggle with the dilemma of federally determined milk prices that don't allow New England dairy farmers to make back their cost of production. As at Adams Farm, the Hunts welcomed the 2013 solar project as one strategy for keeping their farm economically viable.

Red Apple Farm has chosen a very different approach, already discussed in an earlier chapter. The Rose family energetically pursues agritourism at their orchard in Phillipston while exporting local flavor—and nostalgia for old-time New England farms—to two very different venues elsewhere in the state. Skiers at Mount Wachusett in Princeton enjoy Red Apple's hot cider and homemade fudge at the ski lodge and base camp on the mountain, while the smell of fresh-made cider donuts permeates the new Boston Public Market in downtown Boston. Farmer Al Rose has long recognized that the small farms of the Quabbin region remain precarious without some connection to more distant markets. Red Apple's presence outside the area is not about expansion for its own sake but about keeping the Phillipston farm strong and healthy as a part of the local economy and environment.

Another well-known farm on the local scene also thrives because of connections that go far beyond north central Massachusetts. The Farm School follows a scenic ridgetop between Athol and North Orange and incorporates a number of older properties that in many ways still resemble the modest mixed farms of the eighteenth and nineteenth centuries. But the Farm School, founded in 1989, very much reflects the present moment when the average age of American farmers has been rising for decades. Along with camps and farm visits for kids and a vegetable share program with a Boston-area customer base, the Farm School trains young and beginning farmers from around the US, all of them part of the current back-to-the-land generation. Tyson Neukirch, profiled earlier, is one of the Farm School's teachers.

Like Quabbin Harvest Coop, new farmers trying to find land and create businesses that will support them face immense challenges. And yet many of those challenges are very old, as the stories in this book have shown. And as they've always done, farmers are meeting them with dogged ingenuity and often a strong spirit of collaboration. Quabbin-area farmers like Al Rose and Kate Stillman were among the planners and first vendors at the Boston Public Market, created to be a year-round source for foods produced and processed in Massachusetts. Land trusts like Mount Grace Land Conservation Trust and East Quabbin Land Trust are taking on partnerships that link land stewardship with active support for the local food economy. Experienced farmers are teaching and mentoring newer ones. Growing numbers of people in the Quabbin and far beyond continue to engage with big and daunting questions by making changes in how they eat. Over time, those changes can help to shift the balance in our food system and reinvigorate the communities we care about.

a timeline for Quabbin area farming history

pre-1700

Before Europeans settled in the area in the eighteenth century, indigenous groups including the Pequoiag and Nipmuc hunted and gathered a wide variety of food as well as growing crops, especially in the fertile river floodplains.

1700s

European settlers and their descendants created versatile small farms (typically of from fifty to a hundred acres) that included a mix of tilled fields, hayfields, pastures, and woodlots, with an emphasis on pasture-based animal husbandry.

Unlike the more fertile Connecticut River Valley, the area that is now north central Massachusetts was—and remains—a patchwork of different soils and terrains with relatively limited areas of prime soils. By the 1790s, farmable land was in scarce supply even in upland or hill towns like Warwick and Petersham.

The Hardwick Fair started in 1762, the first agricultural fair in what was then still a British colony.

Shays' Rebellion (1786-87) was an unsuccessful armed resistance led by western and central Massachusetts farmers who resented new tax burdens created by the state's Revolutionary War debt, on which financial speculators were demanding payment. As the national economy developed, small farmers began to experience the volatility as well as the new opportunities to be found in more distant markets.

early 1800s

With the "market revolution" of the early nineteenth century, small-scale producers began a long process of adaptation and struggle in an agricultural economy that favored those with the capital to keep up with quickly-changing demands.

Those who didn't own land increasingly found themselves working as wage laborers on others' farms.

New transportation systems like canals and railroads made it possible to ship food over longer distances, a process that has continued into the present.

Dairying, haying, and selling specialty foods like maple syrup took on new importance as strategies for area farmers.

Cheese-making became an important business in many towns. At its peak in the 1850s, Hardwick produced more than three hundred thousand pounds of cheese annually.

mid-1800s

New agricultural imports from New York and the midwestern states created both problems and opportunities for New England farmers. In attempting to keep up with the pressures of expanding markets, many farmers cleared land not suited to cultivation, leading to erosion and depleted soils. Faced with competition from non-local cheese makers, many area farmers shifted toward selling liquid milk or butter, where they still had an advantage in being close to their markets in an age before widespread refrigeration. But imported grain from states to the west also let local farmers raise more animals more profitably.

The Hardwick Fair, like others in the area, gave pride of place to oxen and beef cows at mid-nineteenth century. The town of Warwick held large-scale cattle shows in 1860 and 1861.

late 1800s

We often think of the manufacturing economy overriding the agricultural one. But the growth of industrial towns and cities actually expanded the market for local farm products. Factories also offered a wider range of off-farm, year-round jobs for members of farm families, thus contributing to farm household economies. The number of farms and farmers in New England continued to shrink, and some over-exploited land was reforested as part of a growing turn toward land conservation. However, more intensive farming methods actually increased overall regional agricultural production through the nineteenth century, with a peak in 1910.

late 1800s (continued)

Towns like Athol and Orange expanded quickly with the growth of industries.

In the economically unstable decades following the Civil War, farmers banded together in mutual assistance organizations, trying to shape policies and markets that increasingly favored the wealthy and the growing corporations they controlled.

Economic crises in the 1890s prompted the first back-to-the-land movement while nostalgia for rural places led many urban Americans to tour and vacation on farms, setting a pattern for what is now called agritourism. Many farmers, especially women, offered homemade products at farm stands and took in paying summer guests in search of rest and renewal in the countryside.

Food marketing took place through a patchwork system of small specialty stores, market gardeners, dairies, orchards, and butchers. Many people produced at least some of their own food and purchased the rest from mostly local sources.

early 1900s

The fossil-fuel economy began to take hold, with the advent of automobiles, tractors, petroleum-based fertilizers and pesticides, and an expanding road system. But its growth was uneven in this part of the century, with economic depression and the persistence of smaller-scale food and farm networks countering newer technologies and patterns of living until World War II.

In 1912, the Great Atlantic & Pacific Tea Company, better known as A&P, introduced the economy grocery store model and launched an era of chain store expansion.

In 1930, the first all-in-one grocery store, King Kullen, opened in New York City. It heralded a new level of shopping convenience linked with the expansion of car culture and the provision of ample free parking outside of older town centers. These changes made it increasingly difficult for small local businesses to compete in commercial markets.

During the Depression, more people returned to growing food for themselves and their families. Many small farms faltered, but some new farmers, including immigrants from eastern and southern Europe, were able to buy land at low prices. They often became important suppliers of the produce that still helped feed people in towns and cities.

A number of present-day area farms had their roots in the Depression-era return to the land. In Phillipston, Spaulding and Carolyn Rose, both fresh out of college in 1930, built a commercial orchard business that continues as Red Apple Farm. In 1936, Al and Elsie Diemand bought an older farm property in Wendell that also remains a thriving family business today.

mid-1900s

Petroleum continued to fuel expansion after World War II, enabling highways, long-distance refrigerated trucking, and the continued growth of supermarkets, leading to the sharpest decline yet in the profitability and numbers of New England farms.

Dairying was a mainstay for those who stayed on the land. However, the dairy industry began to follow the same path toward consolidation and corporate control that made it hard for small farmers to compete in other agricultural sectors. The North Central Massachusetts Dairymen's Association, founded by area farmers in 1954, tried to protect small dairies' interests as changes became more entrenched.

At least one local dairy farm got started during this period: Chase Hill Farm was founded in the 1950s and sold milk to bulk distributors for three decades.

Adams Farm, another local dairy, found it difficult to make ends meet and added a small slaughterhouse in 1946. By the early 2000s, it had become the largest animal processing canter in the region.

1960s-1970s

A new back-to-the-land movement in the 1960s and 1970s brought many young homesteaders to the area in search of affordable land, small-scale community, and a lifestyle that felt closer to natural rhythms and foods. Few of these newcomers started commercial farming ventures, but many became important advocates for farmland, open-space preservation, and locally based economies, thus reconciling environmentalism and agriculture through the valuing of traditional working landscapes and community character.

As tourism in rural New England continued to grow, some farms, like Hamilton Orchards in New Salem and Red Apple Farm in Phillipston, were able to attract new customers by offering pick-your-own opportunities.

1980s-1990s

The end of the century saw the rise of the chain grocery superstore as well as the globalization of food production and marketing with expansive new trade agreements.

Several mainstays of the local farming community were also founded in these decades, including Many Hands Organic Farm in Barre (1982), the Farm School in Athol (1989), and Seeds of Solidarity Farm in Orange (1996). The first North Quabbin Garlic and Arts Festival in 1998 began a tradition of highlighting the grass-roots creativity and productivity of an area that is still struggling economically after the loss of much of its major industry.

early 2000s

Enthusiasm for local food spurred the growth of farmers markets, agritourism, and direct marketing in the early twenty-first century. Producers, consumers, and planners work to shorten the long-distance food chains created over the past hundred years. Today's small farmers must contend with the formidable economies of scale represented by large regional and national food producers who are able to keep prices low through enormous volumes. But local food producers have formed new alliances with schools, hospitals, land trusts, planners, and others. More people are recognizing the benefits of less energy-intensive ways of farming. Connections to more populous customer bases in the Pioneer Valley and eastern Massachusetts continue to expand the market for Quabbin local food and support the people and farms that produce it.

Quabbin farm and coop recipes
Julie Rawson
Many Hands Organic Farm

Jack's Favorite Salad Dressing

to be served on the daily farm salad
2 cups organic olive oil
1 cup Bragg's apple cider vinegar
1/2 cup organic ketchup
3 tablespoons honey
sea salt to taste
1 tablespoon dry mustard powder
2 teaspoons paprika
some garlic powder

Mix ingredients in a quart jar, screw on a lid, and shake well.

Julie's Pink Root Soup

Start with pork stock. This is made by boiling down the head, tail, and bones of the pig for a few hours in a big pot of water with sea salt and some Bragg's vinegar. Once it is all boiled, cool and separate the meat from the chunky fat and bones and feed them to your dog. Pour off stock into quarts and add a little of the meat from the head and tail to each quart and freeze. Or, you can buy a quart of certified organic pork stock from us at MHOF: www.mhof.net

In the stock boil your favorite roots. I use onions, beets, carrots, potatoes and sea salt to taste—I use Celtic Sea Salt. When they are soft, cool enough to run through a Vitamix or strong food processor (the pork meat strands can be daunting to a weak machine). Warm slowly and you can add chopped parsley or chives or your favorite green herb on top right before serving. You can also add a dollop of sour cream if you like. The beets give this soup its name.

Jeannette Fellows
Chase Hill Farm

Spinach Feta Stuffed Burgers

1/3 cup crumbled Chase Hill Farm Feta cheese
1 tablespoon minced fresh oregano
2 tablespoons chopped fresh spinach
1 pound Chase Hill Farm organic 100% grassfed ground beef
salt and freshly ground black pepper to taste

Mix feta, spinach, and oregano in a small bowl. Divide the ground beef into 3 equal portions and form into burgers. Place 1/3 of the feta mixture in the middle of each patty and reform into a patty, making sure the stuffing is covered with meat. Sprinkle with salt and pepper and cook or grill until done.

Mac and Blue

8 oz elbow pasta, cooked al dente and drained
3/4 cup walnuts, toasted
1 1/2 cups grated Chase Hill Farm Farmstead cheese
1/2 cup crumbled Chase Hill Farm Quabbin Blue
2 cups whole milk
1/3 cup butter
1/3 cup flour
1 teaspoon kosher salt

Make a white sauce: Melt the butter, whisk in flour, and cook 1 minute. Pour in milk slowly while whisking to combine and cook over medium heat stirring continuously until sauce thickens and becomes silky.

Remove from heat; stir in cheeses until barely melted; slowly add cooked pasta and heat gently. Serve topped with toasted walnuts. Winter comfort food!

Skillet Macaroni and Cheese

2 tablespoons butter
1 cup chopped onion
1/2 teaspoon pepper
1/2 teaspoon oregano
1 teaspoon salt
1 teaspoon dry mustard
1/4 teaspoon thyme
2 cups dry macaroni
3 cups water
3 tablespoons flour
1 1/2 cups milk
2 cups grated Chase Hill Farm Cheddar Cheese (or 1 cup Cheddar and 1 cup Farmstead)

In a large skillet melt butter. Add onion and sauté until tender. Stir in pepper, oregano, salt, mustard, and thyme. Add dry macaroni and water. Cover and simmer for 20 minutes, stirring occasionally. Blend in flour, milk, and cheese. Simmer for 5 minutes until sauce is thickened and cheese is melted.

Cristina Garcia, chef
The Farm School

These recipes were developed by Farm School chef Cristina Garcia. The dishes are made with ingredients that are part of the Quabbin Harvest Coop Basics program, which offers pantry staples at very affordable prices.

Moo-Shu Vegetable BASICS

Moo Shu

1 tablespoon toasted sesame oil
3 tablespoons plus 1 teaspoon canola oil
4 eggs, lightly beaten
3 garlic cloves, finely chopped
    (2 1/2 tablespoons)
1 inch piece of ginger, finely grated
    (1 1/2 tablespoons packed)
1 small onion, thinly sliced (1/2 cup)
1 medium head cabbage, thinly sliced (14 cups)
1 large carrot, grated (2 cups)
2 teaspoons arrowroot powder
6 tablespoons Tamari or soy sauce
2 tablespoons rice vinegar

Warm a wide and shallow pan over low heat with 1 teaspoon canola oil. Add egg and cook gently. Once the thin omelet is mostly set lift edges to let uncooked egg flow under cooked layer. Remove when firm and cool for a bit before rolling and slicing thinly. Set aside.

Whisk together arrowroot powder, Tamari, and rice vinegar.

To keep everything crisp, the Moo Shu will come together in two batches. Overcrowding the pan would steam the cabbage.

Warm half the remaining oils in a wok or large shallow pan over medium high heat. No real need to measure here, just coat the bottom with a thin layer. When oil is hot add half of your aromatics (onion, garlic and ginger), allow to sizzle without browning until fragrant. Add half of the cabbage and carrots. Stir occasionally, you're trying to let steam escape (over stirring here will stew the cabbage). When cabbage and carrots have started to wilt add half of your arrowroot mixture, stir as the sauce thickens. Remove first batch to your serving bowl, tent with foil and repeat. Mix in egg just before serving.

Hoisin Sauce

1 tablespoon canola oil
2 garlic cloves, pasted*
small pinch of ground cloves, cinnamon and
    black pepper
1/2 cup sweet white miso
1/2 cup molasses
5 tablespoons rice vinegar
1/2 cup water

Whisk all ingredients vigorously until smooth.

*To paste garlic, peel and smash each clove. Sprinkle with a generous amount of salt and work into a paste with a mortar and pestle. If you don't have a mortar and pestle, hold a chef's knife almost flat to a cutting board, blade facing away from you. Keep your dominant hand on the handle and the other pinching the top of the blade. With short strokes scrape and press the blade over the garlic, continue until a smooth paste.

Sauce will keep for months in the refrigerator.

Wrap filling in warmed flour tortillas, 1-2 per person or serve over brown rice. Top with Hoisin sauce.

Sriracha Tuna Rolls BASICS
Brown Rice Sriracha Tuna Rolls

Use this technique to roll up other favorites including sweet potato, avocado, cucumber, or carrots. Makes a great school lunch!

1 cup short grain brown rice
1 3/4 cups water
1 tablespoon rice vinegar
2 teaspoons sugar
1 can tuna
2 1/2 tablespoons mayonnaise
2 teaspoons Sriracha
1/2 teaspoon salt
3 nori seaweed sheets, cut in half
1 teaspoon wasabi powder
   (mix according to package directions)
pickled ginger for serving
Tamari for serving

Sushi mat and plastic wrap (optional)

Combine rice and water in a pot. Bring to a boil, reduce to a simmer, and cover with a tight fitting lid. Cook for 45 minutes Remove from the heat, let steam with the lid on for 10 minutes.

Whisk sugar into rice vinegar. Stir into rice, let cool briefly. In another bowl combine tuna and next three ingredients, mix well.

Wrap sushi mat in plastic wrap. Place half sheet of nori, rough side down, at the bottom of sushi mat or clean surface. With wet fingers, spread thin layer of rice (approximately 3 tablespoons) across nori. Leave 1/2" border at the top and bottom.

Place 2 tablespoons of tuna close to the bottom third of the rice, with or without the mat, begin rolling the nori up and over the tuna. Gently squeeze the bamboo mat to secure the filling (be careful if rolling without mat to dry your hands completely before compressing gently).

Once roll is secure, use a clean wet knife to cut roll in half and then your two halves in thirds so you end up with 6 pieces. Serve with wasabi, pickled ginger, and Tamari or soy sauce.

Maple Pudding BASICS

Delicious with a dollop of whipped cream, a shake of maple sugar and a drizzle of maple syrup

2 cups whole milk
1/3 cup maple syrup
2 large egg yolks
3-4 tablespoons cornstarch (depending on
   desired thickness)
1/8 teaspoon salt
2 tablespoons butter
1/2 teaspoon vanilla extract (optional)

In a medium bowl, whisk together milk, maple syrup and yolks. In a medium heavy-bottomed pot, combine salt and cornstarch. Add liquids to pot while whisking. Over medium-low heat, whisk constantly until pudding coats the back of a spoon, 10-12 minutes.

Whisk in butter and optional vanilla. Transfer to a bowl and press plastic wrap to pudding surface to prevent skin from forming. Chill for 2-3 hours. Whisk to smooth before serving.

(The risk with only occasional whisking or a very thin pot is scrambling bits of yolk. If your pudding contains these lumps, pour through a fine mesh strainer before chilling.)

Nina Wellen
founding member of the board, Quabbin Harvest Coop

Squash Fritters

This recipe is a traditional Italian recipe handed down from my grandmother, Frances Cifarelli. She often used the zucchini flowers in the pancakes too.

- 3 medium zucchini, yellow summer squash, or a combination of both
- 3 eggs
- 6 sprigs of flat Italian parsley or chopped basil
- pepper
- vegetable oil
- pancake mix (I use organic Maple Valley or you can make your own)
- 1/4 cup Parmesan cheese, grated

Wash the zucchini, then grate into a bowl. Sprinkle with a little salt and let stand 15 minutes. Squeeze out the excess water and drain. Add enough pancake flour to absorb any remaining moisture. Beat the eggs, add the parsley, pepper, and cheese. Add to the grated zucchini mixture. Stir until well mixed. Add oil to frying pan, about 1/4 inch deep. Spoon batter into small pancake-sized patties. Cook until golden brown on both sides. Remove each fritter with a slotted spatula. Put on a layer of paper towels to drain.

You may need to add a bit more pancake flour as you go, but don't add too much as the fritters will become too "cakey."

You can easily double this recipe. These fritters freeze nicely.

Homemade Tomato Sauce

This sauce recipe has been adapted from the one my grandmother, Frances Cifarelli, passed down to my mother, Sarah Wellen, who passed it down to her children.

- 5 pounds Italian plum tomatoes- you can order these from Quabbin Harvest in late August
- 1 large chopped onion
- 1/2 cup fresh basil, chopped
- 1/4 cup fresh Italian flat parsley, chopped
- 2 cloves garlic- chopped or mashed
- salt to taste
- pepper to taste
- 1/2 cup water
- red pepper flakes to taste

Wash tomatoes, cut in half, remove any blemishes, and place in large pot with chopped onion and the water. Cover and bring to a slow boil. Stir to keep tomatoes from sticking to bottom of the pot. When tomatoes are soft, remove from stove and let cool a bit. Remove seeds and skin by processing cooked tomatoes through the smallest holes of a food mill set over another large pot. Put back on stove and simmer uncovered to a slow boil. Let sauce simmer on low for about a half hour. Add chopped parsley, basil, garlic, red pepper flakes, salt, and pepper. Simmer slowly without a cover for another hour. Taste it, and if tart, add a pinch of sugar. Simmer sauce until fairly thick, stirring often, another hour or so. The lower and slower you simmer, the better the taste and the thicker it gets. I have let sauce simmer 3 or 4 hours. You can add oregano if you want- fresh is best.

When sauce is done you can put it aside and use on pasta later, or after it cools, you can freeze it.

I have used cherry tomatoes in my sauce, and heirloom or yellow tomatoes also; just use a little less water.

Anne Diemand Bucci
Diemand Farm

Finnish Pancakes
Courtesy of Massachusetts Poultry Association, Inc.

- 8 large eggs
- 1 quart fresh milk
- 3/4 tablespoon sugar
- 1 teaspoon salt
- 1 cup flour
- 1/4 pound butter

Spray pan with cooking spray. Melt 1/4 pound butter in 12" x 16" pan. Mix milk and eggs thoroughly with beater. Add sugar, salt, and flour. Pour mixed batter over melted butter.

Bake in 450 degree oven for 20-23 minutes.

Cut recipe in half for 9" x 13" pan. Cut in thirds for 8" x 8" pan.

Serve topped with maple syrup, jelly, or butter.

Pickled Eggs
Courtesy of Massachusetts Poultry Association, Inc.

- 2 cups white vinegar
- 2 tablespoon sugar
- 1 medium onion, sliced
- 1 teaspoon salt
- 1 teaspoon mixed pickling spices
- 12 peeled, hard-cooked eggs

Simmer all ingredients (except eggs) in sauce pan for 10 minutes. Put eggs in jar and cover with hot mixture.

Refrigerate for several hours.

Laura Moore
Moore's Maple Grove Farm

Moore Family Breakfast Cake
Passed down through the family from Great Grandma Moore, who baked them in cast iron muffin pans.
(You can make them in any muffin tins.)

- 2 cups flour
- 1/4 cup oil
- 1/3 cup sugar
- 1 egg
- 1 cup milk
- 3 teaspoons baking powder
- 1/2 teaspoon salt
- 1/2 teaspoon ground nutmeg

Mix all ingredients. Pour into greased muffin tins.

Bake at 375 for 20-25 minutes.

a note on sources

The stories in this book draw on many different kinds of sources: in-person interviews, local histories, old deeds and census records, and scholarly studies of farming and food systems in our area and elsewhere. If you want to learn more, you may want to explore some of these suggestions for further reading and learning.

In any Quabbin area town you'll find a library with a local history collection. **Town histories** can be invaluable sources of information about specific people, farms, and products. In the chapter on Chase Hill Farm, for example, I made use of Charles Morse's *Warwick, Massachusetts: Biography of a Town* (Cambridge: Dresser, Chapman, and Grimes, 1963), which gives a sense not only of how farming in Warwick changed in the time period when the Fellows family have been farming but also how people felt about those changes.

Allen Young's *North of Quabbin Revisited* is another wonderfully detailed source that includes a good deal of information about specific food- and farm-related places in the nine North Quabbin towns as well as great thumbnail sketches of each town and many past and present aspects of life here.

Most area towns are also blessed with active **historical societies** with their own collections and seasonal museums. You'll find fascinating tidbits relating to food and farming there. For example, the Hardwick Historical Society displays artifacts from the town's Grange chapter. Exhibits in the Old Academy Building in New Salem highlight the New Salem Academy's important role in local agricultural education. You'll generally need to do further research to understand the larger contexts for these kinds of fragmentary histories, but there's a richness of detail in these very local sources that's hard to match.

The famed **dioramas at the Fisher Museum of the Harvard Forest in Petersham** are among the best-known sources relating to agricultural change in this part of New England. The dioramas show the same Petersham farm over two centuries, tracing effects of land clearing and reforestation. Viewing the dioramas gives a vivid three-dimensional sense of those changes.

However, there's a downside to the dioramas' vividness. They show us a very straightforward story of farmers clearing too much land that couldn't profitably sustain agricultural production followed by a gradual return to forest. In a nutshell, that's the old familiar narrative about agriculture in New England, and the realistic-looking Harvard Forest dioramas have played no small part in shaping that narrative.

So if you do visit the dioramas, bear in mind that they were constructed in the 1930s in a period when conservationists had come to think that letting older farmland return to forest was the wisest response to ecological and economic crises in the region and the nation. As the chapter on the Neukirch-Anderson Farm in Petersham shows, more recent thinking sees working farms as providing important economic, ecological, and cultural benefits.

That shift in thinking is linked with **recent (and not-so-recent) historical scholarship** that gives a more complex sense of how New England's agricultural landscapes have changed over time. The work of Brian Donahue is very important here; see, for example, his article about the Harvard Forest dioramas, "Another Look from Sanderson's Farm: A Perspective on New England Environmental History and Conservation" (*Environmental History*, Vol. 12, January 2007, pp 9-34), which offers a counter reading to the "decline" narrative.

Hal Barron has written about upland Vermont farming towns that are similar in many ways to those of the Quabbin area; his 1984 book *Those Who Stayed Behind: Rural Society in Nineteenth-century New England* (Cambridge and New York: Cambridge University Press) also questions the narrative of decline. Barron suggests that nineteenth-century New England farming was maturing rather than stagnating and that it suffered by comparison with the then-surging industrial sector.

Michael M. Bell has raised similar questions in articles like "Did New England Go Downhill?" (*Geographical Review*, Vol. 79, No. 4 (October 1989) and "Stone Age New England: A Geology of Morals" in *Creating the Countryside: The Politics of Rural and Environmental Discourse*, edited by E. Melanie DuPuis and Peter Vandergeest (Philadelphia: Temple University Press, 1996).

Other important scholarly works that focus more specifically on our area include the incredibly detailed *Landscape and Material Life in Franklin County, Massachusetts, 1770-1860* by J. Ritchie Garrison (Knoxville: University of Tennessee Press, 1991). Because forested and farmed landscapes have been interwoven here for a very long time, *Stepping Back to Look Forward: A History of the Massachusetts Forest*, edited by Charles H.W. Foster (Petersham, MA: Harvard Forest, 1998) is also very useful.

Two other areas of scholarship help to broaden the context for thinking about local farming and food. If you're looking for good **general sources on the history of agriculture in the US**, the work of R. Douglas Hurt provides a very useful introduction. *American Agriculture: A Brief History* (Ames, Iowa: Iowa State University Press, 1994) is a broad overview, with attention to specific regions including the northeast. Hurt's *Problems of Plenty: The American Farmer in the Twentieth Century* (Chicago: Ivan R. Dee, 2002) focuses more specifically on how farmers have continually tried to keep up with markets that demand efficiencies in ways that outstrip ecological capacities. Raj Patel's *Stuffed and Starved: The Hidden Battle for the World Food System* (Brooklyn: Melville House, 2012) expands this story to the global scale and brings it up to the present day.

Finally, **scholarship about the history of resisting and rethinking large-scale food systems** helps us to understand the cycles of going "back to the land" and the search for alternatives. Dona Brown's *Back to the Land: The Enduring Dream of Self-Sufficiency in Modern America* (Madison: University of Wisconsin Press, 2011) is a fascinating exploration of the American impulse to return to the countryside, especially in the early years of the twentieth century. Warren Belasco's excellent study *Appetite for Change: How the Counterculture Took on the Food Industry* (Ithaca: Cornell University Press, 1993[1989]) updates the story told in Brown's book. Belasco shows how the new homesteaders of the 1960s and 1970s added environmentalism to their other reasons for going "back to the land."

acknowledgments

The Farm Values project and this book owe debts of gratitude to many people, too numerous to name here individually. But I do particularly want to acknowledge the participation of the inspiring and enlightening farmers profiled in the project and the book (and to note that direct quotes attributed to them in the text are taken from interviews conducted during the summer of 2015). Our many supporters at Mount Grace Land Conservation Trust have included Leigh Youngblood, David Graham-Wolf, and Dave Kotker. Also at Mount Grace, Jamie Pottern has been a true star and a good friend. Mass Humanities funded Farm Values with a generous grant. Oliver Scott Snure has been a joy to work with and has created a real gift to the Quabbin local food scene with his beautiful portraits. Marcia Gagliardi at Haley's continues to enrich the life of our area through her deep engagement with publishing its many stories. And among those at the Quabbin Harvest Coop who have been involved in supporting this and other local food efforts, Robin Shtulman deserves special credit for gathering the recipes included in the book as well as for pointing out that we had nearly all the ingredients for a book already in hand!

Writer and scholar Cathy Stanton supports the area's local food economy.

about the author

Cathy Stanton is a writer and scholar who has lived in the Quabbin area for many years. She teaches anthropology at Tufts University and has written about historical reenactment, industrial history museums, and, most recently, the history of food and farming in New England. With Michelle Moon, she is co-author of *Public History and the Food Movement: Adding the Missing Ingredient* (Routledge, 2017). Born in Canada and currently living in Wendell, she serves on the Quabbin Harvest board of directors and is an enthusiastic supporter of the area's local food economy.

about the photographs

Oliver Scott Snure took most of the photographs in A *Quabbin Farm Album*. Based in Northampton and trained at the Hallmark Institute of Photography, he has been photographing people for about a decade. His subjects have included artists, musicians, food, and music festivals. He began photographing Quabbin area farmers in 2015 in cooperation with Quabbin Harvest Coop and the Farm Values project on which this book is largely based.

photo credits

cover, clockwise from top left, Many Hands Organic Farm, Barre, photo by Oliver Scott Snure; haying at the Turkey Farm in 1942, Hardwick, photo courtesy of Susan Twarog; Tyson Neukirch at Neukirch-Anderson Farm, Petersham, photo by Oliver Scott Snure; Fellows family children and cows in 1944, Warwick, photo courtesy of the Fellows family; John Moore at Moore's Maple Grove Farm, Orange, photo by Pamela Moore; Nora Comerford, Athol, date unknown, photo courtesy of Noreen Heath-Paniagua

page 3
flowers at Sweetwater Farm, Petersham, photo by Oliver Scott Snure

page 4
Diemand Farm, Wendell, photo by Oliver Scott Snure

page 6
Ricky Baruc, Seeds of Solidarity Farm, Orange, photo by Oliver Scott Snure

page 7
Tyson Neukirch's hands, Neukirch-Anderson Farm, Petersham, photo by Oliver Scott Snure

page 9
Tyson Neukirch, Neukirch-Anderson Farm, Petersham, photo by Oliver Scott Snure

page 11
Tyson Neukirch, Neukirch-Anderson Farm, Petersham, photo by Oliver Scott Snure

page 12
Adams Farm, Athol, photo by Cathy Stanton

page 13
Nora Comerford, Athol, date unknown, photo courtesy of Noreen Heath-Paniagua

page 14
Nora Comerford and Hester (Comerford) Adams, Athol, date unknown, photo courtesy of Noreen Heath-Paniagua

page 16
at Adams Farm, solar panels overlooking downtown Athol and the Quabbin region, photo by David Brothers

page 17
Noreen Heath-Paniagua, Beverly (Adams) Mundell, Chelsea Frost, photo by Cathy Stanton

PHOTO CREDITS

page 18
John Moore's hand at Moore's Maple Grove Farm, Orange, photo by Oliver Scott Snure

page 19
horses and sledge, Moore's Maple Grove Farm, Orange, date unknown, photo courtesy of the Moore family

page 21
Laura Moore, Maple Grove Farmhouse Bakery, Orange, 2015, photo by Cathy Stanton

page 22
Sienna and John Moore haying, photo courtesy of the Moore family

page 23
John Moore, Moore's Maple Grove Farm, Orange, photo by Oliver Scott Snure

page 24
products of Stillman Quality Meats, Hardwick, photo by Cathy Stanton
young turkeys at Stillman Quality Meats, Hardwick, photo by Cathy Stanton

page 25
Mahan family farm, Hardwick, 1888, photo courtesy of Susan Twarog

page 27
George Crombie and heifer, c 1930s, photo courtesy of Susan Twarog

page 28
John Mahan and George Crombie haying, 1942, photo courtesy of Susan Twarog

page 30
Kate Stillman and helper, Boston Public Market, 2015, photo by Cathy Stanton

page 31
Julie Rawson's hands, Many Hands Organic Farm, Barre, photo by Oliver Scott Snure

page 32
Julie Rawson, Many Hands Organic Farm, Barre, photo by Oliver Scott Snure

page 35
Jack Kittredge and Julie Rawson, Many Hands Organic Farm, Barre, photo by Oliver Scott Snure

page 36
Many Hands Organic Farm, Barre, photo by Oliver Scott Snure

page 37
Mark Fellows's hand, Chase Hill Farm, Warwick, photo by Oliver Scott Snure

page 38
Fellows children and cows, Warwick, 1944, photo courtesy of the Fellows family

page 39
Vida Fellows and children, Warwick, 1935, photo courtesy of the Fellows family

page 40
Oliver Fellows on tractor, mid-1950s, photo courtesy of the Fellows family

page 41
Jeannette and Mark Fellows, Chase Hill Farm, photo by Oliver Scott Snure

page 42
cows grazing, Chase Hill Farm, Warwick, photo by Cathy Stanton

page 44
Mark Fellows and draft horses, Chase Hill Farm, Warwick, photo by Oliver Scott Snure

page 45
Quabbin Harvest Coop, Orange, photo by Cathy Stanton

page 47
Peter Diemand, Anne Diemand Bucci, and Faith Diemand, Diemand Farm, Wendell, from left, top, photo by Oliver Scott Snure; then clockwise, Steve Vieira, Angela Baglione, Lisa Kalan, Laura Davis, and Karen Davis, Sweetwater Farm, Petersham, photo by Oliver Scott Snure; Ricky Baruc, Seeds of Solidarity Farm, Orange, photo by Oliver Scott Snure; Rachel Scherer, The Little White Goat Dairy, Orange, photo by Oliver Scott Snure

page 49
Hunt Farm, Orange, photo by Cathy Stanton

page 50
Red Apple Farm, Phillipston, photo by Cathy Stanton

page 52
Ricky Baruc's hand, Seeds of Solidarity Farm, Orange, photo by Oliver Scott Snure

page 57
sign at Seeds of Solidarity Farm, Orange, photo by Cathy Stanton

page 63
sunflowers at Sweetwater Farm, Petersham, photo by Cathy Stanton

page 67
Stillman Quality Meats, Boston Public Market, photo by Cathy Stanton

Page 69
author Cathy Stanton, photo by Cathy Stanton

page 71
Jack Kittredge's hand, Many Hands Organic Farm, Barre, photo by Oliver Scott Snure

page 73
young pigs, Stillman Quality Meats, Hardwick, photo by Cathy Stanton

> the quick red fox jumped over the lazy brown dog • the quick red fox jumped over the lazy brown dog • the quick red fox jumped over the lazy brown dog • the quick red fox jumped over the lazy brown dog • the quick red fox jumped over the lazy brown dog• the quick red fox jumped

colophon

Text for *A Quabbin Farm Album* is set in Goudy Old Style, also known as simply Goudy, a classic old-style serif typeface originally created by Frederic W. Goudy for American Type Founders (ATF) in 1915.

Suitable for both text and display applications, Goudy Old Style is a graceful, balanced design with a few eccentricities, including the upward-curved ear on the g and the diamond shape of the dots of the i, j; the points found in the period, colon, and exclamation point; and the sharply canted hyphen. The uppercase italic Q has a strong calligraphic quality. Generally classified as a Garalde or sometimes called Aldine face, certain of its attributes—most notably the gently curved, rounded serifs of certain glyphs—suggest a Venetian influence.

Headings for *A Quabbin Family Album* are set in Antique Olive, a humanist sans-serif typeface. Antique is equivalent to sans-serif in French typographic conventions. Along the lines of Gill Sans, Antique Olive was designed in the early 1960s by French typographer Roger Excoffon, an art director and former consultant to the Marseilles-based Fonderie Olive.

In addition to a basic weight, Antique Olive was produced in medium, condensed, wide, bold, condensed bold, extra bold known as Antique Olive Compact, and ultra bold known as Nord. The key shapes, especially the letter O, resemble an olive, one of the unique characteristics of Excoffon's typefaces.

CPSIA information can be obtained
at www.ICGtesting.com
Printed in the USA
FSOW04n0122171217
41987FS